# ELEMENTARY

Laboratory Notebook
3rd Edition

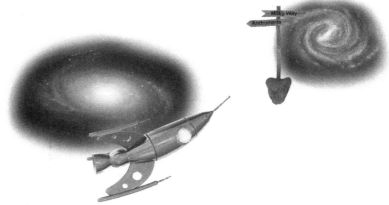

Rebecca W. Keller, PhD

Real Science-4-Kids

Illustrations: Janet Moneymaker

Focus On Elementary Astronomy Laboratory Notebook—3rd Edition
ISBN 978-1-941181-31-7

Published by Gravitas Publications Inc.
www.gravitaspublications.com
www.realscience4kids.com

GRAVITAS
PUBLICATIONS

A Note From the Author

Hi!

In this curriculum you are going to learn how to use the scientific method to explore the world and the universe that you are part of.

In astronomy, making good observations is very important because most of the objects we can see are too far away to visit in person. We learn about them by using a variety of scientific tools and techniques.

Each experiment in this workbook is divided into several different sections. There is a section called *Observe It* where you will make observations. In the *Think About It* section you will answer questions, and in the *What Did You Discover?* section you will write down or draw what you observed in the experiment. There is a section called *Why?* where you will learn about why you may have observed certain things. And finally, there is a section called *Just For Fun* that has an extra activity for you to experiment with.

These experiments will help you learn how to use the scientific method and . . . they're lots of fun!

Enjoy!

Rebecca W. Keller, PhD

Contents

Experiment 1

Observing the Stars

I. Think About It

❶ Think about how you travel from one place to another place.

❷ How do you travel? By road? By plane? By boat? Write or draw your thoughts below.

_____

_____

_____

_____

_____

car

❸ What tools do the driver, pilot, or captain use to navigate? Write or draw below.

_____

_____

_____

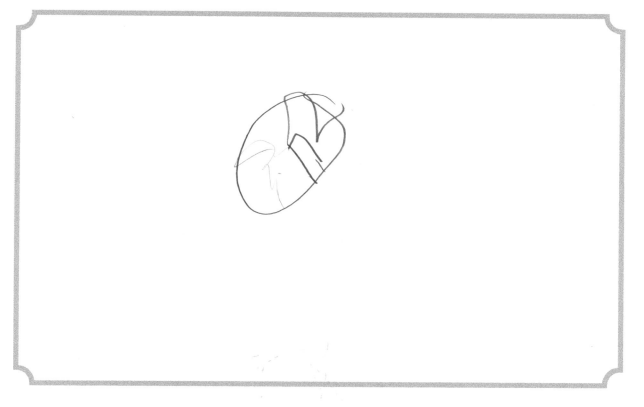

❹ Do you think you can use the stars to navigate? Why or why not?

If you find the ynorth star you can navigate.

☆☆☼◐○☆☆☼☼○☆☆☆☼○☆☆☼◐○☆☆☼☼○☆☆☼○☆☆☼◐○☆☆☆☼○☆☆☼◐○☆☆☼○☆☆☼◐○☆☆

II. Observe It

❶ On a clear night, go outside and observe the stars and the Moon.

❷ In the space below, write down or draw where you are, the time of night, and the direction you are facing.

_____

_____

_____

_____

_____

❸ Draw the stars you observe. Notice bright stars, big stars, and colored stars. Locate the Moon and draw what it looks like. Note stars that are near the Moon and draw them.

NIGHT 1

❹ For the next 5 days, go to the same location and face the same direction as on Night 1. Observe the stars and Moon at the same time each night. Draw what you see. Note if the location of the Moon or stars changes.

NIGHT 2

NIGHT 3

NIGHT 4

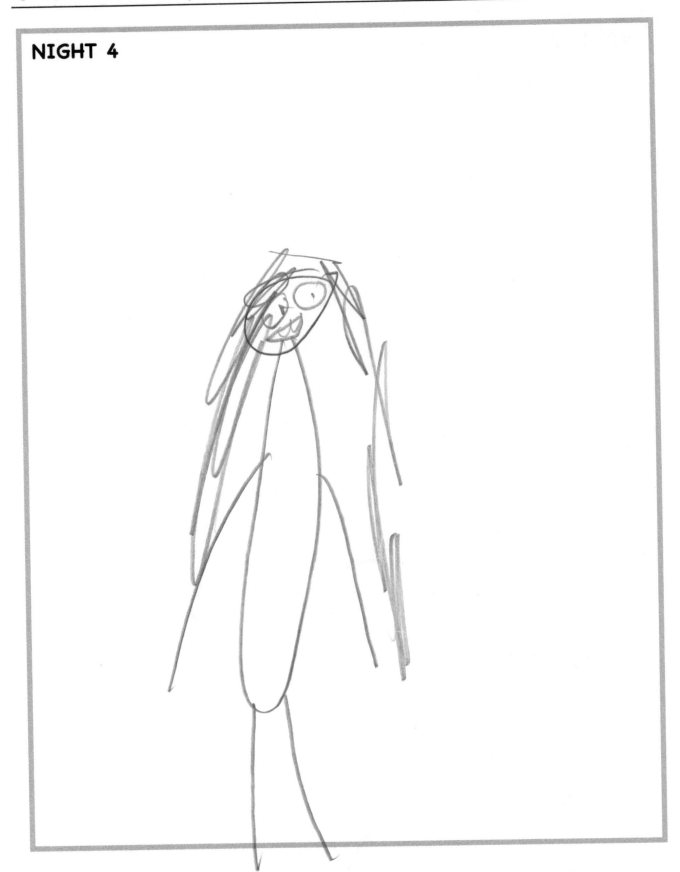

NIGHT 5

NIGHT 6

III. What Did You Discover?

❶ Did any of the stars stay in the same place each night? How do you know?

_____

_____

_____

❷ Did any of the stars change places over several nights? How do you know?

_____

_____

_____

❸ Did the Moon look the same each night? Why or why not?

_____

_____

_____

❹ Did the stars near the Moon stay in the same place each night? Why or why not?

_____

_____

_____

IV. Why?

Astronomers study the stars, planets, Moon, and Sun. Even before astronomy became a science, ancient people used the stars for navigation when they traveled. They also used the stars to plan for changes in weather and even to predict when the Sun would not shine because of a solar eclipse.

For example, the North Star, called Polaris, is above the North Pole and stays in the same place night after night. Using the North Star, travelers can know in which direction they are going. This helps them navigate their journey from one place to another.

Using the stars for navigation takes time to learn because it requires many observations over many days and nights. However, once you know how to use the stars, you can navigate a journey at night without getting lost.

V. Just For Fun

Ancient people observed that some stars look like they are in a group. Stars that appear to be in a group are called a *constellation*. The ancient people gave these constellations names like Orion the Hunter, The Little Dipper, and The Dragon.

On a clear, dark night, observe the stars and find some star groups. What does each of your star groups make you think of? A person? An animal? An imaginary creature? An object? Pick one or more of your constellations and give each a name. Draw the stars in the constellation and connect the stars with lines to show the shape of your constellation.

Constellation _____

Constellation _____

Experiment 2

Building a Telescope

Introduction

In this experiment you will build a telescope and explore how it changes what you can observe.

I. Think About It

❶ Take the sheet of heavy paper, roll it into a tube and tape it together. Next, carefully examine all of the pieces of the telescope.

❷ Which parts are the lenses? How can you tell?

_____

_____

_____

❸ How many tubes are there?

_____

❹ What do you think the different tubes do?

_____

_____

_____

❺ How far do you think you will be able to see with your telescope?

_____

_____

_____

II. Observe It

Assemble and experiment with the telescope.

❶ Take the eyepiece lens and the tube that you made from the heavy paper. Adjust the tube so that it fits around the eyepiece lens. Carefully tape the edges of the lens to the tube. Try not to cover too much of the lens with tape.

❷ Tape the other lens (the objective lens) carefully to one end of the paper towel tube. Again, try not to cover too much of the lens with tape.

❸ Slide the open end of the heavy paper tube that has the eyepiece lens into the open end of the paper towel tube. Your telescope is now ready to use.

❹ In the daylight look at a faraway object with your eyes, then through the telescope. Does the object look different through the telescope?

Try sliding the tube in and out and observe what happens. Does what you are seeing change?

❺ In the daylight observe several different faraway objects, first with only your eyes and then with your telescope. In the spaces provided, draw what you see.

Eyes Only	Telescope

Eyes Only	Telescope

Eyes Only	Telescope

❻ In the evening, once the Sun has set and you can see stars, use your telescope to observe several stars. Draw or describe what you observe. Remember to be patient and make careful observations.

Object 1

Object 2

Object 3

Object 4

Object 5

Object 6

III. What Did You Discover?

❶ How easy or difficult was it to assemble the telescope?

_____

_____

_____

❷ How well did your telescope work?

_____

_____

_____

❸ Were you able to see more details of objects when you used your telescope than with your eyes only? Why or why not?

_____

_____

_____

❹ What features of stars were you able to observe that you couldn't observe with your eyes only?

_____

_____

_____

IV. Why?

Telescopes are tools that make faraway objects appear closer. Astronomers use telescopes to see faraway planets and stars. A basic telescope is easy to assemble and only requires two lenses and a long tube.

By comparing what you see when using only your eyes to what you see when using your telescope, you can better understand just how much a telescope helps astronomers see faraway objects. Part of learning how to make good observations is knowing what something looks like with and without the use of an instrument or tool, such as a telescope. With good observation skills you can see details in the Moon and stars that you might otherwise not notice.

V. Just For Fun

Hold a basketball and carefully observe it. What features can you notice?

Now place the basketball far away and look at it with only your eyes. What features can you see now?

Leaving the basketball in the same place, look at it through your telescope. What features can you see with the telescope that you couldn't see when you used only your eyes?

Try this experiment with the basketball placed at different distances away. You can also try it with other objects. Note your observations in the box provided.

Observations of an Object Near and Far

Experiment 3

Earth in Space

I. Observe It

❶ Cut out the continents on the following page.

❷ Glue or tape the continents onto a large basketball with North America, South America, and Greenland on one side and Australia, Africa, Europe, Russia, and Asia on the other side.

❸ Cut a 2.5 cm (one inch) wide piece from the end of a toilet paper tube. This will give you a nice ring to place the basketball on. When you place the basketball on this cardboard ring, tilt the ball slightly off-center.

❹ Turn off the room lights. Walk several feet away from the basketball and shine light from a flashlight on the basketball.

❺ Leaving the flashlight shining on the basketball, rotate the basketball counterclockwise. Record your observations below.

_____

_____

_____

_____

_____

_____

_____

_____

_____

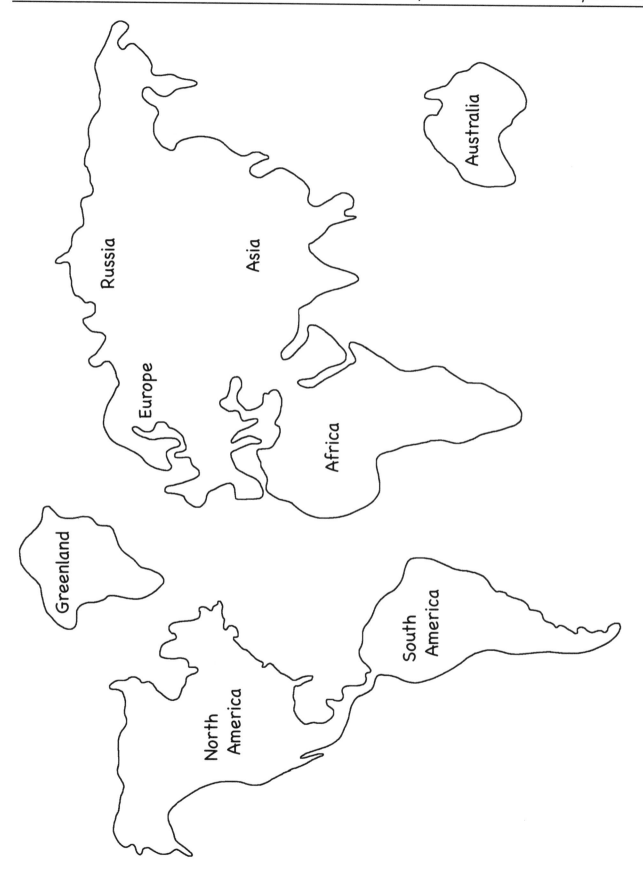

❻ Hold a ping-pong ball a short distance away from the basketball. Move the ping-pong ball in a counterclockwise circle around the basketball. Record your observations below.

_____

_____

_____

_____

_____

_____

II. Think About It

❶ Can you determine how day and night are created by the rotation of Earth?

_____

_____

_____

_____

_____

❷ Can you observe how a lunar eclipse forms (where Earth casts a shadow on the Moon)?

_____

_____

_____

_____

_____

❸ Can you observe how a solar eclipse forms (where the Moon casts a shadow on Earth)?

_____

_____

_____

_____

_____

❹ Using the basketball and flashlight, can you show how the seasons are created? Explain how you would do this.

_____

_____

_____

_____

_____

_____

III. What Did You Discover?

❶ Explain how day and night occur.

_____

_____

_____

_____

❷ Explain how a lunar eclipse occurs.

_____

_____

_____

_____

❸ Explain how a solar eclipse occurs.

_____

_____

_____

_____

❹ What causes the different seasons?

_____

_____

_____

_____

IV. Why?

In this experiment you observed what happens when the Sun shines on the Earth and the Moon. In this experiment the Sun is represented by the flashlight, Earth is represented by the basketball, and the Moon is represented by the ping-pong ball.

When you rotated the basketball (Earth), the flashlight (Sun) was shining on different parts of the ball. This action, (the Sun shining on a rotating Earth) is what causes night and day.

When you took the ping-pong ball (Moon) and rotated it around the basketball (Earth), you observed how the Moon casts a shadow on Earth when the Moon is between the Sun and the Earth. This illustrates a solar eclipse. You also observed how the Earth casts a shadow on the Moon when the Earth is between the Sun and the Moon. This arrangement illustrates a lunar eclipse.

You also found out how seasons occur. The Earth's tilt causes the seasons. As the Earth circles the Sun, some parts of Earth are tilted toward the Sun, receiving more heat energy, and some parts are tilted away, receiving less heat energy. This tilting of the Earth creates seasons as different parts of Earth are tilted toward or away from the Sun.

V. Just For Fun

☼ Think about what it would be like if the Earth's axis still had the same tilt but the axis went through the equator instead of going through the North and South Poles. Would this change the seasons? Would it change night and day?

Try using the basketball and the flashlight to experiment with this idea.

Observations

☼ Next, think about what would be different if the Earth's axis still went through the North and South Poles but was pointed directly at the Sun. Would this change the seasons where you live? Would it change night and day?

Try using the basketball and the flashlight to experiment with this idea.

Observations

Experiment 4

Seeing the Moon

I. Observe It

❶ For fourteen days, observe the Moon at night. Notice any details about how it appears to you.

❷ Record your observations. Note the color and shape.

1	2
3	4

5

6

7

8

9

10

11

12

13

14

II. Think About It

❶ Does the color of the Moon stay the same or does it change?

_____

_____

_____

_____

❷ Does the shape of the Moon stay the same or does it change?

_____

_____

_____

_____

❸ What do you find most interesting about the Moon?

_____

_____

_____

_____

_____

_____

III. What Did You Discover?

❶ Why does the shape of the Moon change?

_____

_____

_____

_____

_____

❷ Where does the Moon get its light?

_____

_____

_____

_____

_____

❸ Do you think the Moon actually changes shape, or does it just look like it has changed shape because the Sun shines on different parts of the Moon on different days?

_____

_____

_____

_____

IV. Why?

In this experiment you observed the shape and color of the Moon for several days. As the Moon circles the Earth, the shape of the Moon appears to change. Depending on when you began observing the Moon, you may have seen a full Moon (completely round), a half Moon (half-round), or a crescent Moon (a curved shape). Or you may not have been able to see the Moon at all.

As the Moon circles the Earth, the Sun illuminates different sections of the part of the Moon that faces us. This makes the shape of the Moon appear to change as we view it from Earth.

Since it takes almost one month for the Moon to circle the Earth, the Moon will cycle through the different shapes each month. The Moon will go from full Moon, to half Moon, to new (dark) Moon, back to half Moon, and then the cycle will begin again.

V. Just For Fun

In the evening find the Moon in the sky and draw what you see. Do you see details on the Moon? Is it a full Moon? Is it a half Moon? Does it look like there's a face on the Moon? Color the Moon as you see it and as you imagine it. Include any details you observe.

THE MOON

Experiment 5

Modeling the Planets

I. Observe It

❶ In this experiment you will explore building models of the planets. Building models is important because it gives scientists a way to help them think about things that they cannot observe close up.

❷ Look at the illustrations of the planets in the *Student Textbook*. Note the sizes and colors of the eight planets. Make notes about what you observe.

_____

_____

_____

_____

_____

_____

_____

_____

_____

_____

❸ Take the eight Styrofoam balls and assign a Styrofoam ball to represent each planet.

❹ Using the information you've collected, paint each Styrofoam ball to look like the planet it represents.

II. Think About It

❶ How did you decide which Styrofoam ball to assign to Jupiter?

_____

_____

_____

❷ How did you decide which Styrofoam ball to assign to Mercury?

_____

_____

_____

❸ What similarities and differences did you notice between the planets?

_____

_____

_____

_____

_____

_____

_____

_____

III. What Did You Discover?

❶ What did building models of the planets help you observe?

_____

_____

_____

_____

❷ What features did you notice that make the planets different from each other? How did you use these features so each planet could be identified from your model of it?

_____

_____

_____

_____

_____

_____

❸ How easy or difficult was it to model the planets?

_____

_____

_____

_____

_____

IV. Why?

In this experiment you explored building models to help you understand more about Earth and the other seven planets that orbit the Sun. Since scientists are not able to go to each of the planets, they use tools to make observations. Then, based on their observations, the scientists make models that show what they think the planets are like.

Models may not be accurate, but they are a scientist's best guess based on the information available. Sometimes using a model will lead a scientist to ask more questions. The answers may add to the scientist's knowledge and can result in changes to the model that make it more accurate.

Scientists can also make mental models of ideas that they have about how things work or why things are the way they are. These mental models may be written in words or pictures or explained with mathematics, and they can lead to many new discoveries.

V. Just For Fun

Think of some other things you could use to build models of the planets. Maybe you could find fruits of different sizes to be the planets. Are there vegetables that would work? Candies? Could you use a mixture of different kinds of items? See if you can invent some different ways to model the planets.

Ideas for Making Planet Models

_____

_____

_____

_____

_____

_____

_____

_____

_____

_____

_____

_____

Planet Model Drawing

Experiment 6

Tracking a Constellation

Introduction

Do you think the stars are always in the same position when you look at the sky? Make some observations to find out.

I. Think About It

❶ How many constellations do you think you could find in the night sky?

_____

❷ Do you think they would change position over the course of a single night? Why or why not?

_____

_____

_____

❸ If you were standing on the North Pole could you see the constellations in the Southern Hemisphere? Why or why not?

_____

_____

_____

❹ If you were standing on the South Pole could you see the Northern Hemisphere constellations? Why or why not?

_____

_____

_____

II. Observe It

❶ Pick your favorite constellation to observe for one week.

❷ On the first night of the experiment, go outside and look for your constellation. If you don't see it the first time out, you can go outside for several nights in a row until you find it.

❸ Note the time, day, and month you first see your constellation.

Constellation _____

Time _____

Day _____

Month _____

❹ Once you find your constellation, record where in the sky you found it. Was it directly above you? Lower in the sky? Towards the east or west?

Position of the Constellation

❺ Follow your constellation by observing it at the same time every day for six more days. Note if it changes location. Record your observations in the following boxes.

Observations   **Date** _____

Observations   **Date** _____

Observations   **Date** _____

Observations   **Date** _____

Observations   **Date** _____

Observations   **Date** _____

Notes _____

_____

III. What Did You Discover?

❶ What is the name of your favorite constellation?

_____

❷ What was the day, time, and month that you first observed your favorite constellation?

_____

_____

_____

❸ What was the season? (summer, spring, fall, or winter?)

_____

❹ How easy or difficult was it to find your constellation? Why?

_____

_____

_____

❺ How easy or difficult was it to follow your constellation for a week? Why?

_____

_____

_____

❻ During that week, did your constellation move?

_____

_____

IV. Why?

Seasonal Movement of Constellations

In this experiment you observed your favorite constellation at the same time each night. You might have discovered that the constellation's position in the sky changed from one night to the next. This occurs because as the Earth orbits the Sun, Earth's position relative to the constellations changes. Each night you are seeing the constellations from a slightly different location in space. The night sky changes during the course of the year as Earth orbits the Sun and changes its position in space.

Try this experiment. Wait a month. Then go outside at the same time you did when viewing the constellation for a week. Look for the constellation. How much has it moved?

Nightly Movement of Constellations

With the exception of Polaris, the North Star, all of the stars and constellations appear to move in the sky during the course of a single night. This is due to Earth's rotation on its axis. As the place where you are located moves around Earth's axis during the night, you see the constellations from different angles. This makes it appear that the constellations are moving through the sky, but actually you are moving. This is similar to the way the Sun appears to move around the Earth. It is really the Earth moving around its axis that causes the Sun's change of position above a particular location on Earth's surface.

V. Just For Fun

Constellations go back to ancient times when people looked at the night sky and noticed groups of stars that reminded them of the shapes of people, animals, and other objects.

For this experiment look at clouds in the sky in the daytime. What do their shapes look like? You may want to lie on your back in the grass while performing this experiment. If there aren't any clouds around, look for a textured surface (like a wall or rock cliff) and see what shapes you can find there.

Once you've found a shape that interests you, draw a picture of it and write a short story about it.

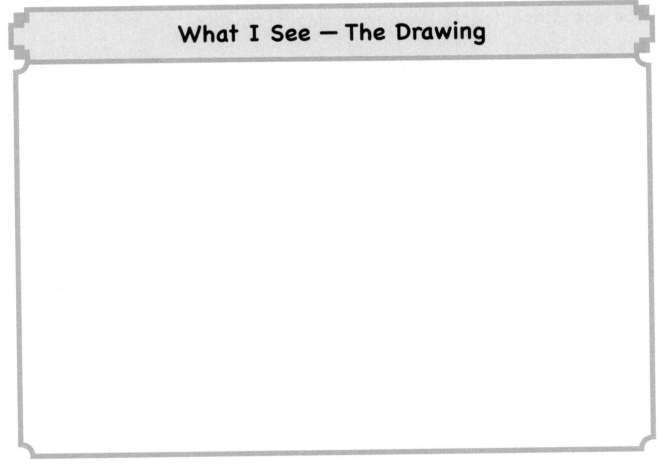

What I See — The Drawing

What I See — The Story

_____

_____

_____

_____

_____

_____

_____

_____

_____

_____

_____

_____

_____

_____

_____

_____

_____

Experiment 7

Modeling an Orbit

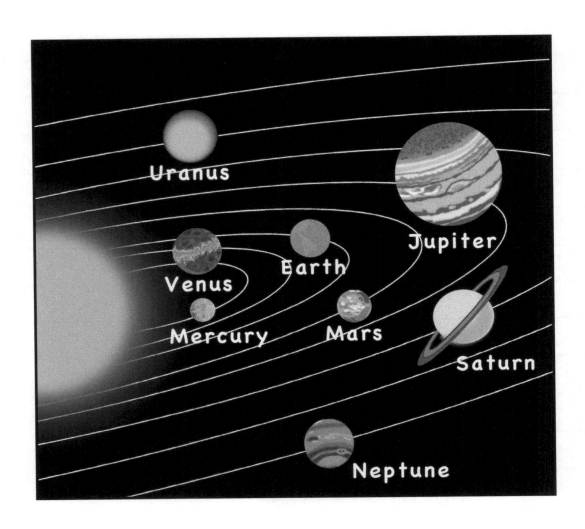

Introduction

In this experiment you will model an orbit by using a Styrofoam ball attached to a string. While holding the end of the string, you will whirl the ball around in a circle. Before you do the experiment, think about what might happen.

I. Think About It

❶ What do you think will happen when you hold the end of the string and whirl the ball?

_____

_____

❷ Do you think the ball will fall towards your hand when you whirl it? Why?

_____

_____

❸ Do you think the ball will fly off the end of the string when you whirl it? Why?

_____

_____

❹ If you shorten the string, do you think the ball will move faster or slower?

_____

_____

II. Observe It

❶ Take a Styrofoam ball, and with the help of an adult, punch a hole through the center.

❷ Next, create a large knot at one end of a piece of string. Thread the unknotted end of the string through the hole in the Styrofoam ball, and pull the string through the ball until you have just enough string to hold firmly in your hand. The end of the string that has the knot will hang down from the other side of the ball.

❸ Hold the unknotted end of the string. The ball should be near your hand. Whirl the string until the Styrofoam ball is moving in a circle with your hand at the center. Observe how the ball moves. In the space below, draw or write about what you see.

❹ Shorten the length of the string by holding it in the middle. The ball will be next to the knot. Whirl the string as in *Step 3*. Observe how the ball moves. In the space below, draw or write what you see.

❺ Shorten the length of the string again, this time holding it close to the Styrofoam ball. Whirl the string as in *Step 3*. Observe how the ball moves. In the space below, draw or write what you see.

III. What Did You Discover?

❶ How easy or difficult was it to use the string to whirl the ball in a circle?

_____

_____

_____

❷ As you were whirling the ball, did it keep moving on a straight path or did it orbit around your hand?

_____

_____

_____

❸ What happened when you shortened the string? Did the ball move slower or faster?

_____

_____

_____

❹ Was it easier or more difficult to whirl the ball with a shorter string? Why?

_____

_____

_____

IV. Why?

When you started whirling the ball, it was near your hand. The whirling motion of the string caused the ball to travel in a circle around your hand and to slide along the string. The ball moved outward away from your hand until it was stopped by the knot. At this point the string was pulling the ball inward toward your hand, and the motion of the ball was pulling it outward, away from your hand. The pulling inward and the pulling outward are different types of force. Once the ball had traveled down the string as far as it could, the inward and outward forces were balanced, and the ball kept traveling in a circular orbit at the same distance from your hand.

The planets move in their orbits around the Sun in much the same way. The Sun's gravity pulls a planet toward it. At the same time, the momentum, or force, of the planet's motion pulls it outward. Because these forces are balanced, each planet stays in a near circular orbit around the Sun.

V. Just For Fun

Place a marble in an empty cup. Now move the cup around in a circle so that the marble travels around the inner surface of the cup. Start moving the cup slowly and then gradually move it faster. What happens as you change the speed? What happens when you move the cup really slowly? What happens if you move the cup in a circle really fast?

Try repeating this experiment using different size marbles and different size cups. Does changing the sizes change your results?

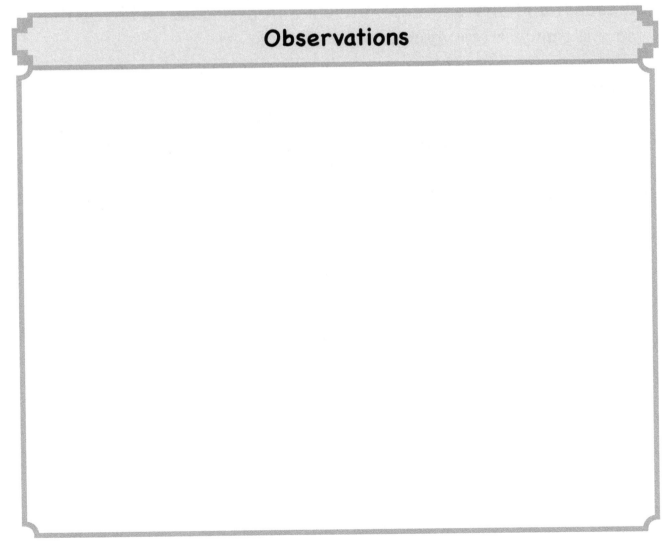

Observations

Observations

Experiment 8

Brightest or Closest?

Introduction

In this experiment you will observe two sources of light: a flashlight and a glow stick.

I. Think About It

❶ Which do you think will give off more light—the flashlight or the glow stick?

_____

❷ Do you think the glow stick or the flashlight will illuminate a path the farthest? Why?

_____

_____

_____

❸ Do you think the flashlight or the glow stick would be easier to see from far away? Why?

_____

_____

_____

❹ Would you use a flashlight or a glow stick to find your way in a dark forest? Why?

_____

_____

_____

II. Observe It

❶ Take the glow stick, and bending it gently, break the inner chamber so the two liquids in it mix and the glow stick lights up.

❷ On a dark night or in a dark room, use the glow stick to illuminate a path in front of you. In the space below record how far you can see.

❸ Take the flashlight and turn it on. Make sure the batteries are fresh.

❹ Repeat *Step 2* with the flashlight. Observe how far you can see with the flashlight. Record your observations below.

❺ Place the shining glow stick and the lit flashlight side-by-side on the ground. Walk several meters (yards) away from them. Being careful not to look directly into the flashlight, observe whether you can see both the flashlight and the glow stick. Record your observations below.

❻ Return to the flashlight and glow stick that are sitting on the ground. Move the flashlight several meters (yards) farther away, behind the glow stick.

❼ Again walk several meters (yards) away from both. Being careful not to look directly into the flashlight, observe the glow stick and flashlight. Record your observations below.

III. What Did You Discover?

❶ Did the glow stick or the flashlight illuminate farther?

_____

❷ With the glow stick and the flashlight side-by-side, could you
   see both clearly? Why or why not?

_____

_____

_____

_____

❸ What happened when you moved the flashlight behind the
   glow stick?

_____

_____

_____

_____

❹ Did the glow stick look brighter than the flashlight when it
   was closer than the flashlight? Why or why not?

_____

_____

_____

_____

☆★☼○★☆☼○★★☆☼○★☆☼○★★☼○★☆☼○★★☼○★★☼○★☆☼○☆★

IV. Why?

How brightly a star shines in the night sky has more to do with how much light energy the star gives out than how close the star is to Earth. A flashlight is able to produce more light energy than a glow stick. You observed this by noticing that a flashlight will illuminate much farther along a path than a glow stick will.

When the glow stick and flashlight are observed from far away, the brightness of the flashlight can overwhelm and wash out the brightness of the glow stick even if the glow stick is closer than the flashlight. The same thing happens with stars. Stars that are brighter but farther away can wash out the light from stars that are less bright but closer. This can make it challenging for astronomers to see dim stars.

V. Just For Fun

Try the same experiment with different colored glow sticks and see if your results change. Does the color matter? Are some colors brighter than others?

Record your observations on the next page.

Observations of Different Colored Glow Sticks

Experiment 9

Modeling a Galaxy

Introduction

Building a model is a great way to learn more about galaxies.

I. Think About It

❶ How many neighborhoods do you have in your city or a city that you have visited?

_____

_____

_____

_____

❷ What else is in the city other than neighborhoods?

_____

_____

_____

_____

❸ What is at the center of a city?

_____

_____

_____

_____

❹ How is a galaxy like a city?

_____

_____

_____

_____

❺ What do you think is at the center of a galaxy?

_____

_____

_____

_____

❻ What do you think is at the center of the universe?

_____

_____

_____

_____

II. Observe It

❶ Making a model involves first thinking about the different features you want to represent and then deciding which materials you will use to represent these features.

Think about the different objects that make up a galaxy. List below the objects you want to represent and the materials you will use to represent them. Do you think all suns and planets look the same as each other?

Objects	Materials

❷ On a clear, flat surface place the materials you chose for modeling a galaxy.

❸ Design your galaxy. Where do the solar system "neighborhoods" go? How far apart are they? What else is in your galaxy? What is in the center? Write and/or draw your ideas below.

Galaxy Model Design

❹ On a piece of cardboard or poster board, create a galaxy as you designed it.

❺ In the box below record any unique features of your galaxy.

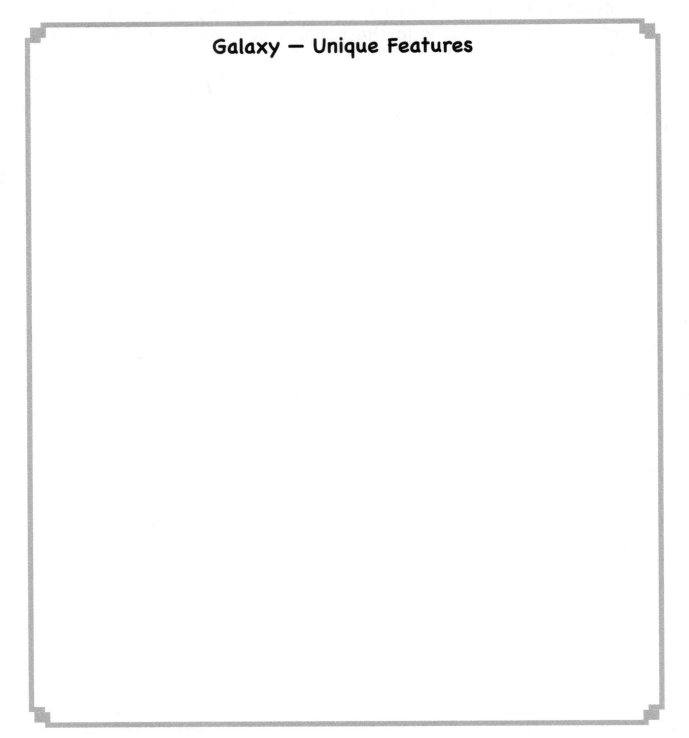

Galaxy — Unique Features

❻ In this box you can tape a photograph you have taken of your galaxy model, draw it, or write about it.

Galaxy Model

III. What Did You Discover?

❶ How many solar systems were you able to put in your galaxy?

_____

❷ How did you decide where to put the solar systems and other objects?

_____

_____

_____

❸ Did you run out of room? Why or why not?

_____

_____

_____

❹ If the universe holds more than 170 billion galaxies, how big do you think it is?

_____

_____

_____

❺ Do you think the universe will ever run out of room? Why or why not?

_____

_____

_____

IV. Why?

Building models is a great way to help us think about things that have different features and many parts. In this experiment you built a galaxy and modeled it. You probably discovered that your galaxy was limited by the size of your cardboard or poster board. The size of the materials you chose to represent the various parts of the galaxy also limited how many of these parts you could include in your model.

We know that cities on Earth run out of room because mountains, rivers, oceans, deserts, and other features of the landscape restrict the amount of space available. Sometimes a city runs out of room because it bumps into the city next to it. This brings up an interesting question that scientists don't know the answer to—does the universe run out of room? Does it have an end? Are there another universes next to it?

V. Just For Fun

Imagine you could travel to the end of the universe. What do you think it might look like? Do you want to include some of the things you saw while getting there?

Draw and/or write your ideas in the following box.

Seeing the End of the Universe

Experiment 10

See the Milky Way

Introduction

In this experiment you will look at the Milky Way.

I. Think About It

❶ Looking at the Milky Way is like looking at your city. If you are near the edge of your city and you can see tall buildings from where you are, look for the area that has the most tall buildings. This will probably be the center of your city.

If you can see the center of your city, draw it below. If you can't actually see the center, draw what you think it might look like. If you don't live in a city, think about one you have visited or seen in movies or pictures and draw what you think a city center looks like.

❷ When you look away from the center of the city you will probably see fewer buildings. If you can look away from the center of your city, draw what you see below. If you can't actually see away from the center or if you don't live in a city, draw what you think the part of the city that is farther away from the center might look like.

❸ Let the buildings in the center of your city represent stars in our galaxy. In the space below, draw what the stars in the center of our galaxy might look like.

❹ Letting the buildings that are farther away from the city center represent stars, draw what you think the stars that are farther away from the center of our galaxy might look like.

II. Observe It

❶ Find an area that is free of city lights on a clear night when there is no moon.

❷ Look into the night sky without a telescope or binoculars. Use only your eyes.

❸ Study the sky and observe areas where there are lots of stars. Compare this to areas with fewer stars.

❹ See if you can find a band of light and stars stretching across the night sky.

❺ In the space below, draw what you observe.

III. What Did You Discover?

❶ How many stars did you see?

_____

_____

❷ Were there areas with lots of stars and other areas with fewer stars? How would you describe the areas of the sky?

_____

_____

_____

❸ Were you able to see a band of stars stretching across the night sky? If so, how would you describe it?

_____

_____

_____

❹ If you could see this band of stars, do you think you were seeing the center of the Milky Way Galaxy, the edge of the Milky Way Galaxy, or something else? Why?

_____

_____

_____

_____

IV. Why?

Earth has an atmosphere that allows us to look through the air to see the stars, Moon, planets, and other objects in our Milky Way Galaxy. Most of the stars and other objects we see in the night sky are part of our galaxy. Earth is located at a perfect spot within the Milky Way Galaxy to observe what surrounds us.

When you see a narrow band of light and stars stretching across the night sky, this band is referred to as the Milky Way. To see this, you are actually looking through the spiral arms of the Milky Way Galaxy toward the center of the galaxy. Because the Milky Way Galaxy is a flat, disk-shaped spiral and we are looking at it edge-on, the stars you observe as you look through the spiral arms appear as a band of light across the sky. If we lived closer to the center of the galaxy, we would see so many stars all around us that it would be difficult to know which way to look to see toward the center.

V. Just For Fun

If you have a computer and would like to see the Milky Way Galaxy, download Google Earth from the internet. Follow the setup instructions. Click on the planet symbol at the top, choose "Sky" from the drop down menu, and type "Milky Way" in the search box.

What did you discover? On the next page write or draw what you found out.

Milky Way Discoveries

Experiment 11

How Do Galaxies Get Their Shape?

Introduction

Galaxies are groups of stars, planets, comets, asteroids, dust, and other things, such as gases. Scientists think that the force of gravity causes all of these objects in space to clump together in groups that have different shapes. Gravity is the force that holds everything together in a galaxy.

I. Think About It

❶ How do you think galaxies form?

_____

_____

_____

_____

_____

_____

❷ How do you think planets and stars are held together?

_____

_____

_____

_____

_____

_____

❸ What do you think causes spiral galaxies?

_____

_____

_____

_____

_____

_____

❹ What do you think causes irregular galaxies?

_____

_____

_____

_____

_____

_____

II. Observe It

In this experiment you will simulate the force of gravity on stars and other objects in space by using a magnet to move small metallic particles. Magnetic force is different from gravitational force but similar enough to use it to model galaxy formation.

❶ Take a shallow, flat-bottomed plastic container and fill it to just below the top with corn syrup.

❷ Pour the iron filings on top of the syrup, being careful to not breathe them in.

❸ Carefully cover the container with plastic wrap.

❹ Place two magnets underneath the plastic container and observe the iron filings. Record your observations in the space below.

❺ Take one of the magnets and create a swirling pattern. Record your observations in the space below.

☆☆☼О☆☆☼ О☆☆☼О☆☆☼ О☆☆☆☼О☆☆☼ О☆☆☼ О☆☆☼О☆☆☆☼О☆☆☼О☆

❻ Take both magnets and create opposite swirling patterns
Record your observations below.

❼ Bring the two magnets together and observe what happens. Record your observations.

❽ Play with the magnets and iron filings. Try moving the magnets in different ways. Record your observations.

Magnet Movement: _____

_____

Magnet Movement: _____

_____

Magnet Movement: _____

_____

III. What Did You Discover?

❶ What happened to the iron filings when you placed the magnets below them?

_____

_____

_____

_____

_____

_____

❷ When you swirled one magnet, did spiral arms form? Was there a center?

_____

_____

_____

_____

_____

_____

❸ What happened when you brought the two magnets together and allowed the iron filings to follow?

_____

_____

_____

_____

_____

_____

❹ Were you able to create any irregular shapes? Describe below what you did.

_____

_____

_____

_____

_____

_____

IV. Why?

Galaxies form because the gravitational forces of stars pull on each other and on planets, comets, asteroids, ice, and dirt. When a force pulls on an object, the object will begin to move.

In this experiment you built a model using iron filings and magnets to observe what is possible when forces move objects. You were able to see how magnetic forces pull on iron filings to create different shapes. In much the same way, the gravitational forces of stars pull on each other and on planets and other objects in space to create the shapes of galaxies.

V. Just For Fun

Make a Jell-O galaxy.

With the help of an adult, follow the instructions on a box of flavored gelatin. Add grapes, berries, or other fruits cut into small pieces. Before the gelatin cools, swirl the fruit into a spiral galaxy, bar galaxy, or irregular galaxy. How many different kinds of galaxies can you make?

In the following box, record your observations

Jell-O Galaxies

Experiment 12

Making a Comet

Introduction

A comet is a mixture of dirt and ice. When a comet travels close enough to the Sun, the ice will vaporize, turning into gas. This gas then creates a tail that follows the comet as it moves through space.

I. Think About It

❶ Draw what you think a comet in space might look like when it is far away from the Sun.

❷ Draw what you think a comet in space might look like as it gets close to the Sun.

II. Observe It

❶ Collect some dirt and small stones.

❷ Pour the dirt and stones into a small pail and cover them with water. Leave several inches between the water and the top of the pail.

❸ Place the pail in the freezer and allow the water to freeze.

❹ Tap the frozen model comet out of the pail.

❺ Observe the frozen model comet. In the space below, draw or write what you see.

☆✧☼○☆✧☼○☆✧☼○☆✧☼○☆✧☼○☆✧☼○☆✧☼○☆✧☼○☆✧☼○☆

❻ Observe the model comet as it melts. Draw or write your observations.

❼ Repeat Steps ❶-❹ using more water. Record your observations.

❽ Repeat Steps ❶-❹ using more dirt. Record your observations.

III. What Did You Discover?

❶ Do you think your frozen mixture of water, dirt, and rocks looks like a real comet? Why or why not?

_____

_____

_____

_____

❷ What happened as your comet melted? Did it come apart in chunks, or did it melt slowly?

_____

_____

_____

_____

❸ How quickly do you think your comet would come apart if it were near the Sun? Why?

_____

_____

_____

_____

❹ How much ice do you think a comet would need to have for you to be able to see its tail from Earth? Why?

_____

_____

_____

_____

_____

❺ How much bigger than your comet model do you think a real comet is? Why?

_____

_____

_____

_____

_____

IV. Why?

In this experiment you modeled a comet. Comets are large chunks of ice and rock that move through space. A real comet might look similar to the small comet model you made from ice, dirt, and rocks, but it would be much larger.

The ice in your comet model melted, but in a real comet the ice would vaporize, turning into gas without becoming a liquid first. Although the method by which the comet loses its ice is different in your experiment than it is for a real comet, this model lets you see what happens to a comet as it loses ice. It gets smaller and begins to break apart until the comet eventually disappears.

Scientists are not always able to make models that work exactly like the object they are modeling. But by making substitutions, scientists can still make valuable observations about objects they cannot get close to.

V. Just For Fun

With the help of an adult, make a water, dirt, and rock mixture and then add dry ice to it. How does the dry ice change your comet?

More REAL SCIENCE-4-KIDS Books
by Rebecca W. Keller, PhD

Building Blocks Series
yearlong study program — each Student Textbook has accompanying Laboratory Notebook, Teacher's Manual, Lesson Plan, Study Notebook, Quizzes, and Graphics Package

Exploring the Building Blocks of Science Book K (Activity Book)
Exploring the Building Blocks of Science Book 1
Exploring the Building Blocks of Science Book 2
Exploring the Building Blocks of Science Book 3
Exploring the Building Blocks of Science Book 4
Exploring the Building Blocks of Science Book 5
Exploring the Building Blocks of Science Book 6
Exploring the Building Blocks of Science Book 7
Exploring the Building Blocks of Science Book 8

Focus Series
unit study program — each title has a Student Textbook with accompanying Laboratory Notebook, Teacher's Manual, Lesson Plan, Study Notebook, Quizzes, and Graphics Package

Focus On Elementary Chemistry
Focus On Elementary Biology
Focus On Elementary Physics
Focus On Elementary Geology
Focus On Elementary Astronomy

Focus On Middle School Chemistry
Focus On Middle School Biology
Focus On Middle School Physics
Focus On Middle School Geology
Focus On Middle School Astronomy

Focus On High School Chemistry

Super Simple Science Experiments

21 Super Simple Chemistry Experiments
21 Super Simple Biology Experiments
21 Super Simple Physics Experiments
21 Super Simple Geology Experiments
21 Super Simple Astronomy Experiments
101 Super Simple Science Experiments

Note: A few titles may still be in production.

Gravitas Publications Inc.
www.gravitaspublications.com
www.realscience4kids.com

GRAVITAS
PUBLICATIONS

CPSIA information can be obtained
at www.ICGtesting.com
Printed in the USA
FFHW011124160120
57757615-63076FF